ISMAEL SADOU

RICE

ISMAEL SADOU

RICE

SUDANO-SAHELIAN ZONE OF CAMEROON (EXTREME NORTH)

ScienciaScripts

Imprint
Any brand names and product names mentioned in this book are subject to trademark, brand or patent protection and are trademarks or registered trademarks of their respective holders. The use of brand names, product names, common names, trade names, product descriptions etc. even without a particular marking in this work is in no way to be construed to mean that such names may be regarded as unrestricted in respect of trademark and brand protection legislation and could thus be used by anyone.

Cover image: www.ingimage.com

This book is a translation from the original published under ISBN 978-613-9-54197-3.

Publisher:
Sciencia Scripts
is a trademark of
Dodo Books Indian Ocean Ltd. and OmniScriptum S.R.L publishing group

120 High Road, East Finchley, London, N2 9ED, United Kingdom
Str. Armeneasca 28/1, office 1, Chisinau MD-2012, Republic of Moldova, Europe
Managing Directors: Ieva Konstantinova, Victoria Ursu
info@omniscriptum.com

Printed at: see last page
ISBN: 978-620-8-51298-9

CONTENTS

INTRODUCTION 2

RICE.. 6

CONCLUSION 36

BIBLIOGRAPHICAL REFERENCES 37

INTRODUCTION

Rice (Oryza sativa L.) is the staple food of more than half the world's population (FAO, 2013). According to the United States Department of Agriculture USDA) (2014), records were set in both global rice production at 465.6 million and consumption at 468.7 million in 2013 (USDA, 2014).

According to the FAO (2013), world rice consumption is set to increase between 2012 and 2023 by just over 1% per year, compared with 1.7% for the 90s and 2010s. Average per capita consumption is expected to rise slightly compared with the reference period, to 58.2 kg per person (FAO, 2013).

In Africa, increasing production can be achieved both increasing the area under cultivation and by improving yields. In Cameroon, national rice production is estimated at 311,674 tonnes of paddy grown on 44,000 ha, covering only 38% of needs (FAO, 2013). As a result, imports have risen from 21,000 tonnes in 1975 to 728,443 tonnes in 2017, despite the country's potential, hence the need to increase production to reduce particularly expensive imports (FAO, 2017).

Production by the Société d'Expansion et de Modernisation de la Riziculture de Yagoua (SEMRY) covers an area of 11,518 hectares, accounting for more than half of irrigated land and providing at least 90% of national production (FAO, 2013). African rice growing is subject to a number of constraints that seriously hamper production. There are two types of obstacle: abiotic obstacles (soil and climate) and biotic obstacles: insects, weeds, diseases, birds, nematodes and rodents (Heinrichs et al., 2004). Insect pests of rice constitute one of the main groups of biological agents limiting rice production in Cameroon (Nacro, 2014). The main insect pest species commonly found on rice fall into several orders (WARDA, 2016). However, according to Nacro (2014), four (4) orders stand out as being the most important. These are Lepidoptera, Isoptera, Coleoptera and

Diptera, which cause considerable damage to rice fields (Vaissayre, 2007). In Cameroon, pest attacks are one of the reasons for low yields. Yield losses can be as high as 30% for rainfed rice and total for lowland rice (Dakouo and Nacro, 2012).

The use of chemical insecticides was favoured to deal with these large and diverse groups of pests (Vaissayre et al., 2006). As soon as they appeared on the scene, synthetic insecticides attracted attention because of their efficacy and the immediate benefits they brought (Crétenet et al., 2015). Generally speaking, the advent of pesticides made it possible to curb the action of pests, and they were seen as a panacea before coming to a halt (Guèye, 2012).

Excessive use of synthetic pesticides has encouraged pests to become accustomed to them and then to develop resistance (Vaissayre et al., 2006). Pesticides have also become a cause for concern in terms of public health, the environment and the loss of biodiversity (Multigner, 2005, Adam et al., 2010).

For all these reasons, stakeholders in the rice sector in Cameroon are increasingly interested in alternative, environmentally friendly control methods such as plant extracts with an insecticidal effect. A great deal of work in this area has demonstrated the effectiveness of plant extracts (El Hadj et al., 2006; Gbedomon et al., 2012; Fayalo et al., 2014; Tamgno et al., 2014). It therefore seemed appropriate to focus this study on plant extracts, including neem. The Meliaceae are among the plants most widely tested for their contact effect, and neem is undoubtedly the species most studied in this context (Facknath, 2006). Products extracted from neem have proved effective against more than 400 species of arthropod crop pests and nematodes (Bélanger and Musabyimana, 2005).Control using biopesticides such as aqueous neem seed extract makes a remarkable contribution to modern environmental science by emphasising the need for balance in nature (Facknath, 2006). It includes all methods that reduce the nuisance of rice pests, excluding chemical pesticides and control methods that directly reduce the populations of pests present on rice. Several studies have assessed the potential of pesticidal plants traditionally used in Africa. Their conclusions are interesting: many plants have a real effect on crop pests (Bélanger and Musabyimana, 2005).

This is particularly true of neem (Azadirachta indica), a tree native to India whose leaves and seeds have insecticidal, antifungal and vermifuge properties (Bélanger and Musabyimana, 2005). Applying neem extracts to tomato crops can reduce the severity of fungal infections, limit the hatching of lepidopteran eggs, and modify fecundity and behaviour of certain insects (Gbedomon et al., 2012). Caterpillar and aphid populations are lower on plots treated in this way than on others. Similar results can be obtained with other plants such as spicy substances extracted from Guinea pepper (Xylopia aethiopica), black mustard (Sinapsis nigra) and tobacco (Nicotiana tabacum) (Facknath, 2006). In plots treated with these extracts, populations of beetles, whiteflies and thrips are reduced by 61 to 78% in size (Facknath, 2006). Eradication is not total, but yields are equivalent to those obtained on plots treated with synthetic insecticides. What's more, these effects can be enhanced by mixing extracts from different pesticidal plant species.

Finally, some species, such as those belonging to the genus Ocimum (which includes basil), the advantage of not only pesticidal properties, also medicinal properties, and can also be eaten as leafy vegetables or spices. This versatility greatly enhances their appeal (Facknath, 2006). Pesticidal plants do not eliminate all pests, but they keep their populations below the nuisance threshold, while offering many advantages over synthetic pesticides (Bélanger and Musabyimana, 2005). Generally less hazardous to health, pesticide plant extracts break down rapidly in This reduces the risk of environmental pollution and improves the health of the produce grown. Cultivated in association, pesticide plants help to maintain the balance between crop pests and associated animal auxiliaries. In some cases, these natural products can be used to increase yields, at a cost-benefit ratio similar to that of synthetic pesticides (Bélanger and Musabyimana, 2005). To realise their full potential, however, a number of barriers still need to be overcome.

The first concerns their acceptance by farmers, who consider that their use is too restrictive (time needed to produce the extracts, number of treatments required, limited specificity of the extracts, variability of results, etc.). Their effects on crop protection agents (ladybirds, for example) are not well known. Their cost is also a problem, because for the moment these products, when they are intended to be marketed, are manufactured in small quantities by small production units (Bélanger and Musabyimana, 2005).

Lastly, they unable to comply with the assessment procedures required by the (rare) existing regulatory frameworks. These are very onerous and are in fact the same as those required for synthetic pesticides, and small production structures cannot comply with them. This constraint in particular limits the commercial viability of pesticide plants. In addition to these intrinsic and structural problems, the development of plant pesticides is also hampered by the availability and dissemination of knowledge. Although the use of pesticide plants is an ancestral practice, the associated knowledge is fragmented and dispersed within communities. The work carried out as part of this thesis therefore proposes alternative methods of protecting rice, including the use of extracts from natural plants such as neem fruit, which have insecticidal properties, are inexpensive and environmentally friendly.

RICE

1.Origin

Rice is a monocotyledonous plant in the Gramineae or Poaceae family (Delseny, 2000). Cultivated rice belongs to two botanical species of the Oryza genus (Oryza sativa and Oryza glaberrima). The most widely cultivated species is Oryza sativa L, which is diploid, self-pollinating and native to South and South-East Asia (Second, 1982). It is thought to have been domesticated from Oryza rufigon in India and China over 10,000 years ago (De Datta, 1981). Europeans introduced Asian rice to Africa from the fifteenth century and later to America. The species Oryza glaberrima or African rice or Casamance rice originated in Central Africa and has been domesticated in West Africa for 3,500 years. A third species of ricc, a cross between Oryza sativa L (Asian rice) and Oryza glaberrima (African rice), is grown in Africa and the rest of the . "New African Rice" or NERICA. Its germplasm offers real hope for improving the productivity, profitability and sustainability of rice growing in Africa (WARDA, 2006).

2.Systematic position

According to Denis Delaval and Wopereiset al., 2008, rice is an annual grass of tropical origin belonging to the :

- Plant kingdom: multi-cellular organism with an underground root system and an above-ground vegetative system;
- Super-branch of cormophytes: plants with stems, leaves and roots;
- Phylum of Spermaphytes or phanerogams: cormophytes flowers and seeds;
- Sub-branch of the Angiosperms: plant in which the ovule is contained in a

closed ovary;

- Class Monocotyledons: seed plants with one cotyledon

- Order: Poales

- Super family: Cereals

- Family: Grasses or Poacea: family of monocotyledonous plants with inconspicuous flowers grouped in spikes.

- Genus Oryzae: annual herbaceous plants with inherited stems with nodes and internodes and leaves borne by nodes on the stem.

- Spccics: Oryzac sativa L and Oryzae glaberrima Steud.

The Oryzae genus around twenty species, only two of which are cultivated (Oryzae sativa L and Oryzae glaberrima Steud) Wopereis et al., 2008.

3.Characteristics of the different rice species

3.1.Oryzae sativa

Oryzae sativa, Asian rice, comprises two main types: indica and japonica.

a) Type indica

The indica type, native tropical Asia, is generally characterised by :

• light-green, long, broad to narrow leaves ;

• significant tillering ;

• grains that are generally long and fine ;

• and many secondary branches (small branches at the level of the panicle).

b) Japonica type

The japonica type, native to the temperate and subtropical zone Asia, has :

• dark green, narrow leaves ;

• average tillering ;

• a short to intermediate size ;

• and grains that are often short and round.

3.2.Oryzae glaberrima

Oryzae glaberrima, the African rice, originates from the central Niger delta. was Steudel who first named and described Oryzae glaberrima on samples in 1855. from the West African coast (Guinea, Portuguese). It can be recognised by the following characteristics:

• excellent vegetative development, leaves and glumes glabrous (smooth, hairless), ligules small and rounded;
• low density of branches few or no secondary branches red caryopsis (grain with a red pericarp) long dormancy.
Today, the Asian species (Oryzae sativa) is much more widely cultivated than the African species (Oryzae glaberrima) (Wopereis et al., 2008).

Photo 1: Comparison of the panicle of Oryzae glaberrima Steud (right) with Oryzae sativa (left) (Wopereis et al., 2008).

4. Plant structure

Rice is a very flexible plant that grows in both flooded and non-flooded areas. The rice plant consists of round, hollow stems, flat leaves and terminal panicles. It comprises the vegetative organs (roots, stems, leaves) and the reproductive organs (i.e. the panicle made up a set spikelets) (Wopereis et al., 2008). Rice grows well in an aquatic environment and likes more water for its growth and development (Audebert, 2015).

5. Vegetative organs

The vegetative organs in rice consist of roots, leaves and stem, which are easily identifiable, and other organs such as the sheath, ligule, auricle, etc. (Wopereis et al., 2008).

a) The roots

The roots are the underground organs of the rice plant that serve as a support (anchor) for the plant. They made up of secondary roots and absorbent hairs through which nutrients are taken from the soil to feed the plant. The primary root, which grows from the seed at the moment of germination, only lives for a short time. It is quickly replaced by secondary roots (Wopereis et al., 2008). According to Marc Lacharme (2001), rice roots are the shallow underground organs whose main functions are to absorb and store water and nutrients.

b) The stem

The rice stem is made up of a series of nodes and internodes, which are hollow with a smooth surface. The lower internodes are shorter than the upper ones (the narrower the distance between the lower nodes, the more resistant the plant is to lodging; the sturdiness of the stems (diameter) and their size are also criteria for resistance to lodging); on the rice plant, each node bears a leaf and a bud which may develop into a tillers. The main functions of the stems are to transport nutrients and water and to supply the roots with air (Wopereis et al. 2008). The nodes of the main stem give rise to other stems called secondary tillers, which may in turn bear tertiary tillers. All the tillers produced by a single plant make up the rice tuft, and tillering depends on the variety but is also influenced by growing conditions and practices. The main stem (primary tillers) develops the largest number of leaves, and at the base of the tillers the first rudimentary leaf is called the prophyllum (Marc Lacharme, 2001).

c) **Leaves**

Leaves are the aerial organs that develop alternately on the stem. They originate at each node of the stem and are made up of two parts, the leaf sheath and the leaf blade. Each node gives rise to a leaf, with the leaf sheath enveloping the entire inter-node area and in some cases the following node. The leaf blade or terminal part of the leaf is attached to the node by the leaf sheath and the last leaf is attached to the node by the leaf sheath. or terminal leaf below the panicle is called a panicle leaf or flag leaf (Marc Lacharme, 2001). The leaf is the engine of growth, capturing solar radiation and transforming it into carbohydrates, allowing the plant to breathe and transpire, and it is in the leaf that the plant carries out photosynthesis to produce the carbon elements (carbohydrates) that are essential for plant metabolism. Rice leaves can be erect, oblique or drooping; this characteristic (leaf habit), which depends on the variety, has a major influence on the penetration of solar radiation (Wopereis et al., 2008).

The sheath is the part of the leaf that surrounds the stem. At the junction between the leaf and the sheath (collar) are two structures called the auricle and ligule.

The auricle is sort of crescent-shaped appendage measuring 2 to 5 mm, covered in hairs.

The ligule is a kind of membrane whose length and shape depend on the species and variety. It is long in Oryzae sativa, but short and rounded in Oryzae glaberrima. Rice is the only grass that has both the ligule and the auricle, which distinguishes it from weeds at the seedling stage (Wopereis et al., 2008).

6. Reproductive organs

The reproductive organ, also known as the floral organ, in rice consists of the panicle, flowers and grains or paddy (Wopereis et al. 2008).

a) The panicle

It is the terminal part of the rice plant and is borne by the last internode. The panicle is made up of primary branches (small branches) which carry secondary branches carrying the pedicels which in turn carry the spikelets. The number of primary and secondary branches depends on the species and variety. A rice panicle can bear between 50 and 500 spikelets, but for most of the varieties used, the number of spikelets is between 150 and 350 and there are varietal differences between the length, shape and angle of the panicles (Wopereis et al. 2008).

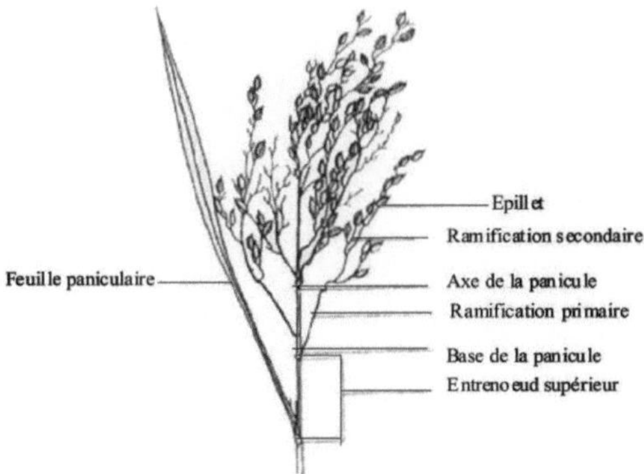

Figure 2: rice (Oryzae sativa) panicle, (Wopereis et al., 2008).

b) The flower

The flower consists of male reproductive organs (anthers containing pollen) and female organs (the ovary). Rice is a self-pollinating plant, fertilisation being provided by pollen from the flower itself, unlike the allogamous plant, which is fertilised by pollen from another flower of the same plant, such as maize (Marc Lacharme, 2001). It is the flower that is responsible for the formation of the grains obtained after fertilisation (Wopereis et al., 2008).

c) Grain or paddy

The rice grain is made up of three essential parts (Wopereis et al. 2008):
- the husk of the rice plant, which comprises: the glumes (large portions above the pedicels that link the spikelets to the secondary branches) and the two glumes known as the palaea (upper glume with three veins) and the lemma (lower glume with five veins). The barb is an extension of the ventral vein of the lower glumella. The glumellae, which envelop the rice grain (caryopsis), make up the rice ;

- the endosperm which serves food source for the embryo ;
- the embryo located on the ventral part of the spikelet, also known as the germ, gives rise to the young seedling after germination.

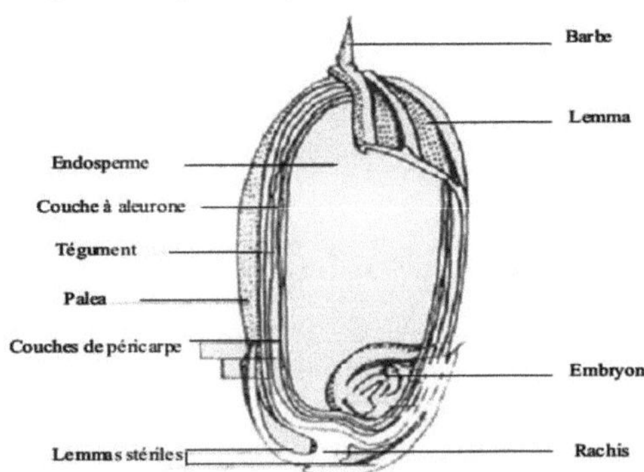

Figure 3: The rice grain (Wopereis et al., 2008).

7.Growth and development stages of rice

rice cycle, whatever the variety or ecology, is complete once it has gone through the following ten development stages (Wopereis et al., 2008).

7.1.Germination

The embryo, which slows down during storage, germinates as soon as it encounters sufficient humidity (the equivalent of a quarter of the seed's weight) and a favourable temperature (optimum: 20 to 35°C). Germination is the return to activity of the seed; emergence is materialised either by the appearance of the coleoptile, from which the first leaf will emerge (aerobic conditions) or by the appearance of the radicle, the first root (anaerobic conditions). The germination stage corresponds to the period between the appearance of the coleoptile or radicle and the emergence of the first leaf (Wopereis et al., 2008). According to Marc Lacharme, 2001, taking temperature into account, the germination phase lasts from 5 to 20 days (5 days in warm conditions and 20 days at low temperatures).

7.2.Seedling

This is the period following germination, during which the young shoot feeds mainly on the seed's reserves (around 14 days). It continues to produce leaves at a rate of one every 3 or 4 days. The seedling stage corresponds to the period between the emergence of the first leaf and the appearance of the fifth leaf. During this stage, the seedling also produces roots, a critical stage during which the plant is very fragile (Wopereis et al., 2008). For Marc Lacharme, in 2001, emergence extends from emergence to the 4-leaf stage and lasts 15 to 25 days depending on the temperature (low temperatures lengthen emergence). During this phase, the plant gradually becomes independent of the grain's food reserves. The plant is completely independent by the 3-leaf stage. If sowing in a nursery, it is necessary to wait until this stage before transplanting. The duration [germination + emergence] is about 21 days for winter sowing. It can be extended to almost 40 days for sowing at the very start of the warm counter-season (Marc Lacharme, 2001).

7.3.Tallage

Tillering is the period when the young seedling begins to produce tillers. The tillering stage begins with the appearance of the fifth leaf and lasts for a variable length of time, depending on climatic conditions (temperature) and the variety. The tillers continue to increase in number until they reach maximum tillering, a period known as the tillering phase. Some tillers then degenerate and the number of tillers stabilises (Wopereis et al., 2008). According to Marc Lacharme, in 2001, a long-cycle variety will have a higher tillering capacity than a short-cycle variety. 3 to 5 days before the end of tillering, panicle initiation can be observed inside the stems of different tillers.

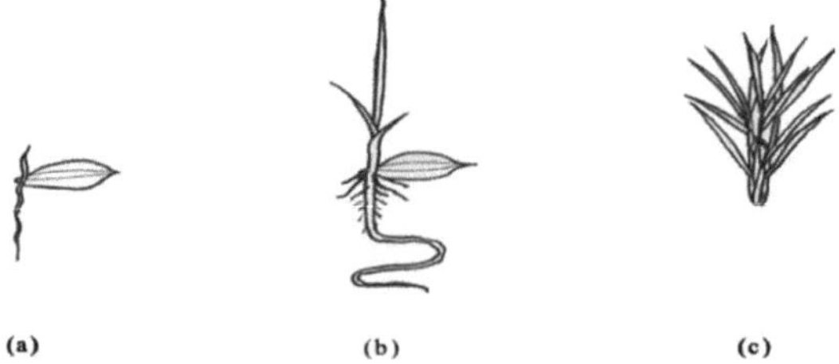

Figure 4: Emergence (a), emergence (b) and tillering (c), from Wopereis et al., 2008.

7.4. Elongation of internodes

Towards the end of the tillering phase, the plant begins to lengthen its internodes. This results in an increase in the size of the plant, so it becomes long enough to reach its size for flower initiation.

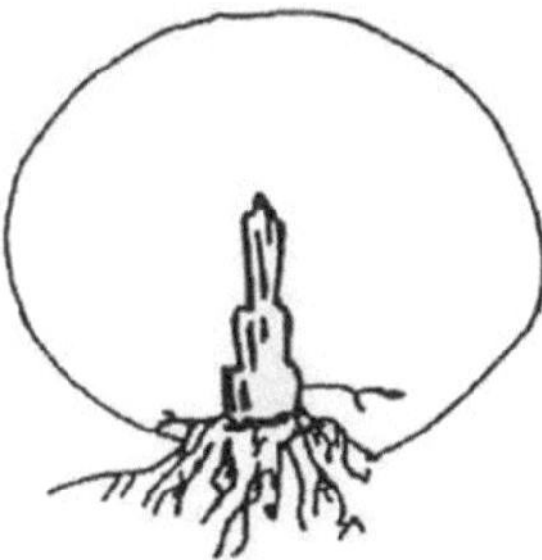

Figure 5: Internode according to Wopereis et al., 2008.

7.5. Panicular initiation

This stage marks the birth of the panicle. The young panicle, which emerges from within and at the base of the last node, is first materialised by a small feathery cone-shaped outgrowth of 1 to 1.5 mm, only visible when the stem is dissected. This stage in the plant's development is known as panicular initiation (PI) and the number of grains per panicle is determined. For early varieties, maximum tillering, internode elongation and panicle initiation occur almost simultaneously. In rice, the date of panicle initiation is determined by several factors, some of which are constants inherent to the variety, temperature and photoperiod (Wopereis et al., 2008). Panicle initiation marks the start of the reproductive phase, which contributes to grain formation.

7.6. Development of the panicle

According to Wopereis et al. 2008, this stage is characterised by swelling at the base of the panicle leaf due to the panicle moving up the stem. After initiation, the panicle develops and progresses up the stem, causing the stem to swell, known as bolting. The various organs of the flower develop and the panicle continues to grow, reaching its final size before emerging (heading).

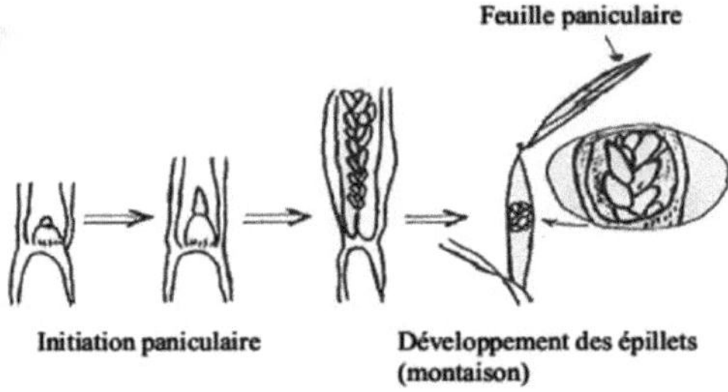

Figure 6: Evolution of panicle development from Wopereis et al., 2008

7.7. Emergence and flowering

Heading (appearance of the ear) is characterised by the emergence of the panicle at the base of the panicle leaf. The spike takes one to two weeks to emerge completely from the stem. Flowering refers to the opening of the flower and subsequent pollination. The flower opens under the pressure of two transparent structures, called lodiculi, located at the base and inside the seed, which expand by heating, at the same time spreading the two glumellae (Lemma and Paléa). In rice, this opening generally occurs between 9 and 11 in the morning. As soon the glumellae open, the stamens stand up. and because of the ambient temperature, the anthers dry out and release the pollen grains, which fall onto the stigmas and then onto the ovary, where fertilisation takes place. The two glumellae then close (Wopereis et al., 2008). Heading is the development of the ear, which generally occurs three days after flowering and continues progressively until the panicle is fully emerged and the grains have developed.

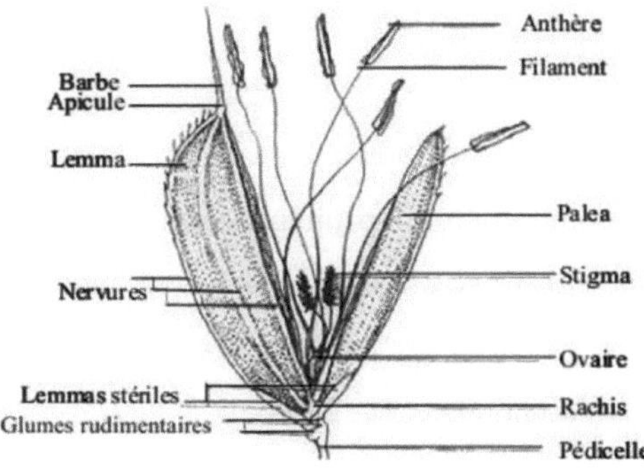

Figure 7: Heading and flowering stages to Wopereis et al., 2008.

7.8. Milky grain

After fertilisation, the ovary swells and the caryopsis develops, reaching its maximum size in around seven days. The contents of the grain (caryopsis) are watery at first, then take on a visible milky consistency. At this stage, the panicles are still green, upright and the plant is no longer growing (Wopereis et al., 2008).

7.9. Pasty grain

At this stage, the milky part of the grain becomes soft and then turns into a hard paste (two weeks after flowering). The panicle begins to bend and the colour of the kernels gradually changes from green to the colour characteristic of the variety (straw yellow, black, etc.), which is obtained when the kernels reach maturity (Wopereis et al., 2008).

7.10.Maturity

It is at this stage that the seeds ripen and reach their final size, their maximum weight is also reached and gives the panicle its curved shape. The grain becomes hard and takes on the definitive colour of the variety (straw yellow, black, etc.). This stage is reached when 85-90% of the grains on the panicle are ripe and ready to be harvested (Wopereis et al., 2008).

8.Plant development phases

The development phases of rice are all the life stages through which rice passes to complete its entire growing cycle. According to Marc Lacharme in 2001, the development stages of rice are grouped into three main phases:

- the vegetative phase, germination to panicle initiation;
- the reproductive phase, panicle initiation to flowering;
- and the ripening phase, which runs from flowering to full ripening of the seeds.

According to WARDA in 1995, the biological cycle of annual rice takes place in several stages. During its growth, rice goes through three essential phases: a vegetative phase, from germination to the initiation of floral primordials; a reproductive phase, from this initiation to pollination; and finally a grain maturation phase. It should therefore be noted that rice is an annual plant and its cultivation goes through three phases (the vegetative phase, the reproductive phase and the maturity phase). The duration of the vegetative phase varies significantly between species and varieties under the same growing conditions, while the reproductive phase remains more or less constant whatever the variety (WARDA, 1995).

8.1. Vegetative phase

During the vegetative phase, the rice plant goes through the following stages of development: germination-emergence, seedling, tillering and internode elongation (depending on the type of variety). The duration of the vegetative phase is variable and depends on the variety. This phase is influenced by low temperatures and the photoperiod (length of the day), which can, when the variety is sensitive, extend its duration. Most cultivation practices are carried out during the vegetative phase, in particular weed control. Weed control (weeding, plant protection treatments), fertilisation, insect and disease control (Marc Lacharme, 2001).

8.2. Reproductive phase

The reproductive phase extends from panicular initiation to fertilisation. It includes the stages of panicle initiation, bolting, heading and fertilisation. It is characterised by the birth of the panicle and the development of the spikelets and reproductive organs. It has a relatively fixed duration of between 30 and 35 days, whatever the variety and the season. The reproductive phase is not affected by photoperiod, although it is very sensitive to low temperatures, drought (water deficit) and salinity, which can cause sterility of the spikelet reproductive organs, resulting empty grains (Wopereis, 2008).

8.3. Maturity phase

This phase runs from fertilisation of the seeds to maturity. It includes the flowering stage, the milky stage, the doughy stage and the ripening stage. It lasts for a relatively fixed period of between 30 and 42 days, whatever the variety, season and environmental temperature and humidity conditions (Marc Lacharme, May 2001). It is sensitive to climatic hazards such as high temperatures, violent winds and drought (water deficit) during the first 15 days after flowering (pasty stage). Draining plots or stopping irrigation at the doughy stage does not have a negative impact on production, but is actually beneficial for the rice; it homogenises ripening and facilitates harvesting (Wopereis, 2008). It is at this stage that the grains reach their final maturity for harvesting, drying for consumption and other purposes.

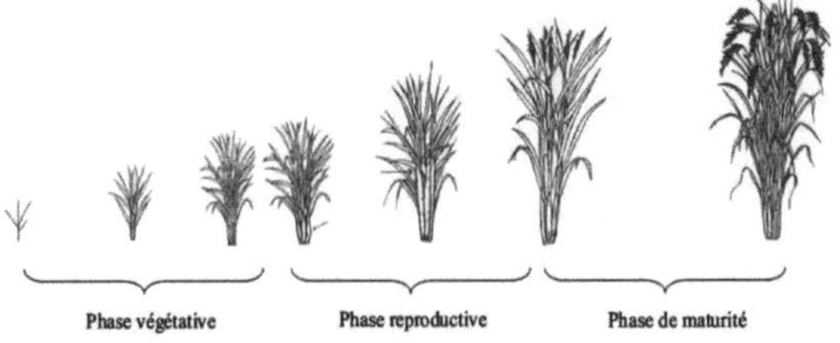

Figure 8: The three main phases in rice development, according to Wopereis et al., 2008

9. Rice ecology

In 2001, Schalbroeck concluded that the best soil and geographical conditions are necessary for good rice production.

9.1. Altitude and humidity requirements

Rice grows at sea level as well as at altitudes of up to 3,000 m (in Nepal), provided that the conditions, especially heat and water, are met (Audebert, 2015). Flowering requires 70 to 80% humidity, "a light wind is favourable to rice," and "winds that are too strong cause damage (Schalbroeck, 2001).

9.2.Soil

Irrigated rice prefers heavy, fine-textured soils with 40-60% clay and low percolation water losses. Loose silt to clay soil (with a higher clay content (50%)) is best suited to rice cultivation. The optimum pH varies between 6 and 7, but irrigated rice can tolerate pH values of 4.5 to 8.5, because after a rice field is submerged, the pH that was acidic increases and the pH that was basic decreases by about two units. Values of 5 and 8 are the limit pH values for suitable soil (Schalbroeck, 2001). For dry cultivation, rice requires rich, loose soil with good water retention capacity and an optimum pH of between 6 and 7(Audebert, 2015).

9.3.Temperature

Below 12-13 °C, germination is impossible and some varieties require a minimum temperature of 18 °C. After transplanting, plants can recover at an average daily temperature of around 13 to 15 °C (Schalbroeck, 2001).

Table 1: Air temperatures required for rice growing

Development stage	Air temperature (oC)		
	Minimum	Optimum	Maximum
Germination	14-16	30-35	42
Tallage	16-18	28-30	40
Flowering	22	27-29	40
Maturation	/	25	40

Source: Schalbroeck (2001)

9.4. Water requirements

Irrigated rice growing with a permanent water table requires a water table of 1,000 to 1,500 mm, taking into account losses in the canals; on filtering soil, this quantity of water can be doubled (WARDA, 1995). In rainfed rice growing, a well-distributed rainfall of 800 to 1000 mm is sufficient for a vegetative cycle of 4 to 5 months. Depending on climatic conditions and the length of the vegetative cycle, water requirements in terms of potential evapotranspiration are between 400 and 800 mm (Schalbroeck, 2001).

The critical periods for water are :

- tillering ;
- heading and bolting;
- maturation.

Water requirements correlate with soil : on clay-loam soils, rice can be grown with only 800 to 1000 mm of water.

9.5. Light

Rice requires a lot of light, and photoperiodism has a strong influence cycle length and yield. Sunshine plays an important role in rice growth and yield by encouraging tillering and increasing the number of spikelets and seed weight. A minimum of 400 hours of sunshine during the first two months of cultivation, including 220 to 240 hours during the last month, is necessary to obtain a high yield (Schalbroeck, 2001).

10.Importance of rice

Rice is a commodity that can be used for a variety of purposes, including human, animal and industrial consumption. It is a source of employment and income, and also plays an important role in other areas.

10.1.food and feed

Rice is used in human nutrition and is the staple food in 17 countries in Asia and the Pacific, 9 countries in South and North America and 8 countries in Africa. It provides 20% of the world's food energy requirements (WARDA, 2002). Rice provides dietary energy (rice is rich in carbohydrates, it is a source of vitamin B, zinc, magnesium and phosphorus) in different regions of the world to improve dietary diversity and provide essential micronutrients. Rice is not only an important source dietary energy, it is also a source of thiamine, riboflavin and niacin. Wholegrain rice contains a significant amount of fibre. The amino acid profile of rice shows a high level of glutamic and aspartic acid, while the limiting amino acid is lysine. However, rice cannot provide all the nutrients required for a healthy diet; it must be accompanied by meat products, fish and pulses (WARDA, 2002). Hulls, straw and crushed bran are used in livestock farming to feed cattle, while crushed or uncrushed grain, broken grain and flour are used on farms to rear poultry.

10.2.Source of income and employment

Rice is a strategic commodity that contributes significantly to food security by providing income and employment for the rural poor. , rice accounts for around 12% of agricultural value (Kline and Gordon, 2012) quoted by Salifou B. DIARRA in 2014. Given the large number of rice growers (two million), local rice is an important element in the fight against poverty in rural areas, as long this activity can be made profitable through development initiatives. If the rice sector is developed, jobs will be created in all segments of the rice value chain, particularly production, services and distribution. Local rice is also a source of economic growth through its effects on the substitution of massive rice imports, generating significant outflows of estimates. The contribution of the rice sector in terms of taxes and duties is relatively significant both upstream and downstream in some countries (WARDA, 2002).

10.3. Other uses

Rice is used in some companies to make alcohol, varnish, oil, vinegar, glucose, acetone and pharmaceutical products. Around the world, particularly in West Africa and Asia, rice is used to perform certain traditional rituals and for medicinal purposes, for example the grains are eaten or swallowed orally to combat diarrhoea. Rice husks are used as fuel and their ashes as fertiliser, while the straw can be used as bedding and raw materials in the manufacture of paper pulp (SEMRY, 2011).

11. Different rice ecosystems

There are two main rice-growing ecosystems depending on the water regime (Trébuil and Houssain, 2004) cited by Audebert et al., (2015):

- aquatic ecosystem, with at least temporary controllable water ; known as a paddy field;
- a non-aquatic ecosystem, where rice, like other cereals, is grown on flooded and drained soils.

11.1. Aquatic rice crops

They cover 88% of rice-growing areas and are subdivided into four categories (Audebert et al., 2015).

11.1.1. Irrigated rice

Hydro-agricultural infrastructures make it possible to control when water enters and leaves the rice field, as well as the height of the water surface. It was in this type of rice growing, which covers 55% of cultivated land and accounts for 75% of world production, that the "green revolution" of the 1960s took place. The simultaneous use of highly productive semi-dwarf varieties, mineral fertilisers and pesticides, combined with good weed control thanks to transplanting and manual weeding, resulted in average yields of 4-5 t/ha and maximum yields of 10 t/ha. Rice monoculture is often the rule, and in certain hot climates, the use of early, non-photosensitive varieties allows up to three crop cycles per year. In subtropical areas, rice-wheat rotation is . With the rise in costs, the trend is to abandon transplanting in favour of sowing.

11.1.2. Flooded rice fields

In this type, there is no hydro-agricultural water control infrastructure, and the water supply to the rice field depends directly on rainfall or river flooding; the dates of arrival and withdrawal of water in the rice field are not controlled, and the water level can vary from 0 to 100 cm. This type of cultivation covers 23% of the world's land area. The most common method of planting the crop is sowing. Yields rarely exceed 4 t/ha. The major concern is that yields are stagnating at around 3 t/ha. The varieties used need to be hardier, and their height and cycle well adapted to the water regime.

11.1.3. Deep submerged rice or "floating" rice

Cultivated in deltaic areas with water levels of up to 5m, it accounts for 10% of cultivated areas and only 3% of world production. The varieties used are characterised by a rapid elongation of the internodes as the water rises. Yields are low (around 1 t/ha) and subject to considerable spatial and inter-annual variability. It is becoming less and less attractive.

11.1.4. Mangrove rice cultivation

It is practised on coastal plains subject to the influence of the tides, and relies on good management of fresh and salt water. During the rainy season, the rice field is desalinated (the salt is removed) by submerging it in fresh water, then the rice is planted and protected from the salt water by dykes. The varieties used must have a good tolerance to salinity. Given the high fertility of the soil, yields of 5 t/ha can be achieved without the use of fertiliser. In rice fields, the existence of anaerobic conditions can lead to the appearance of problems linked to soil characteristics (acidity, salinity, alkalinity, iron toxicity, etc.). The desire to save water resources and the high cost of development mean that good water management is essential.

11.2. Rainfed rice

It is a non-submerged crop, fed by rain and/or groundwater. It accounts for 12% of the world's land area, and 40% in Africa. Rainfed rice is traditionally grown in slash-and-burn systems. These systems are becoming less and less productive as a result of shorter "They also increasingly come up against the need to protect forests and combat erosion. They are also increasingly coming up against concerns about protecting forests and combating erosion. Fixing rain-fed rice cultivation is a major development issue. The examples of some densely populated areas in Africa and some large farms in Brazil show that this is technically possible (Syaukat and Pandey, 2004; Pinheiro et al., 2006), cited by Alain Audebert et al. in 2015.In Cameroon, rice is grown in three ecosystems: flooded, irrigated and rainfed.

12. Constraints on rice production

There are a number of obstacles to rice production in Africa. There are two types of obstacle: abiotic (soil and climate) and biotic (weeds, diseases, birds, nematodes, rodents and insects) (Heinrichs and Barrion, 2004).

12.1. Insect pests of rice

The diversity of their mouthparts, their diets and their ability to adapt to almost any environment make insects the most formidable and devastating of all crop pests. Rice cultivation is therefore faced with a number of problems caused by insect pests, which affect almost every part of the plant from the beginning to the end of the production cycle; for rice alone, some 4098 species of insect pests have been inventoried (Kumar, 1990). Most of these insect pests are Lepidoptera, Diptera, Coleoptera and Hemiptera, and some of them are chewing, chewing-licking, sucking and biting-sucking insects. Depending on their damage or attacks on plants, they can be grouped into five main groups (Wopereis et al., 2008):

- stem borers ;

- defoliators ;

- biters, suckers ;

- root cutters ;

- stock pests.

12.1.1.Stem drills

There are around fifty species divided into two orders: Lepidoptera and Diptera. The most important species in the order Lepidoptera are : Chilo spp, Maliarpha separatella, Sesamia calamistis, Nymphula depunctalis, Eldana saccharina, Scirpophaga spp. The Diptera order includes Diopsis spp and Orseolla oryzivora. The latter species causes rice midge (Wopereis et al., 2008). Stem borers are the most important pests, infesting rice plants from the seedling stage to maturity. There are two main types of damage (Defoer et al., 2004):

✓ Damage caused to young plants at the beginning and during tillering. Here, the caterpillars penetrate the leaf sheaths and the base of the young stem; the damage can lead to the death of the stem.

✓ Damage caused from the panicle. The young caterpillars (especially of the ^{2nd} generation) gather a few centimetres below the panicle inside the flower stalk. This dries out completely, leaving an entirely white or dried-out panicle.

Apart these two highly visible cases, any damage to the stem is also harmful. When the insect attacks a plant at an advanced stage of development, the old caterpillar lodges in the lower parts of the stem and can reduce or stop the panicle from feeding. As a result, one or more branches dry out, reducing the number of harvestable grains. For these attacks by stem borers, the order of Lepidoptera (Sesamia sp) and Diptera (Diopsis thoracica West) are the most represented (Wopereis et al., 2008).

12.1.2. Defoliators

Damage caused by leaf destroyers generally involves removing fragments from the tips of the leaves. In other cases, the leaves may be cut off completely. This damage is often of little economic importance when the plant is sufficiently developed. However, at the start of the crop and up to the middle of tillering, infestations can affect a large part of the plant, leading to irreparable destruction (Wopereis et al., 2008). The order most commonly found in this type of attack is Coleoptera and Orthoptera. In the Coleoptera order we have Nymphula depunctalis, which is the most widespread species for this attack. It is a small white butterfly, 10 to 12 mm long, which deposits its eggs in tight rows along the leaves. As soon as it is born, the larva climbs onto a leaf and shears it off, leaving a flap that serves as an attachment. The leaf fragment curls up but does not swell and closes with a few silk films, finishes cutting the remaining piece of leaf, then drops into water. It swims towards a rice stalk on to which it climbs and feeds. This allows them to move easily from one stem to another. Drying out for at least three days is an effective means of control. Other defoliating species include : Cnaphalocrocis medinalis, Marasmia trapezalis, Diacrisia scortilla, Parnara spp, Hispides spp (Heinrichs and Barrion, 2004).

12.1.3. Stitch-suckers

Biting and sucking insects can attack leaves, stems or seeds. They are sometimes numerous and can cause significant damage in the event of a large outbreak. This category includes bugs, whiteflies, aphids, Cicadellidae and mites, which cause damage mainly during the ripening phase, at the milky and pasty stages. The most represented order in this group is the Hemiptera (Wopereis et al., 2008).

12.1.4. Root cutters

Root cutters are insects, most of which develop partially or totally in the soil. Some are found exclusively in lowland areas, while others are only found in upland conditions (Wopereis et al., 2008).

- The rice moth

These insects belong to the same order as locusts. Although they are present in all ecosystems, crickets are more abundant in uplands and their damage is greatest at the edge of fields.

- Termites

They can be found in low-lying areas as well as on plateaux. Their presence is always favoured by the absence of water. This is why flooding plots is the best way to combat these insects. They feed on the roots and underground parts of the plant, causing the leaves to turn yellow and dry out. The species most commonly found in rice fields are Microtermes, Macrotermes and Trinervitermes spp.

- Water weevil

This insect lives exclusively in the shallows and muddy waters of rice fields. The larva is the harmful stage that feeds exclusively on rice roots.

12.1.4. Stock pests

Stored rice is attacked by a wide range of insects whose identity and cycle differ from those encountered in the field. Most of these are polyphagous insects that attack not only the rice but also other stored grains. Stored grain insects fall into two groups depending on the mode of infestation. Some start attacking crops in the field and are introduced into the granary or shop through the harvest. This is the case with weevils and alucites. Others only attack during storage; they come either from inside or outside the shop (Heinrichs and Barrion, 2004).

12.2. Birds and rodents

12.2.1. Birds

Rice probably suffers more from bird damage than from insect damage. The rice field is a particularly favourable environment for permanent waterfowl and many migratory species (Heinrichs and Barrion, 2004). Permanent waterfowl can nest in rice paddies by crushing the rice clumps, and consume the young or ascending stems by shredding them and leaving a lot of debris (Grist and Lever, 1969. Anon, 1970). As for migratory birds, granivorous passerines (an order of small birds) are the most dangerous. They sometimes swoop down on the rice paddies in considerable numbers as they approach maturity. The damage is spectacular: panicles whose milky grains have been sucked out one by one, leaving whitish traces of dried milk on the glumes, broken stems and panicles, ravaged rice fields lying more or less flat over very large areas (Heinrichs and Barrion, 2004).

12.2.2. Rodents

Rats are particularly active enemies both in the rice field and in the paddy and rice stocks after harvest. In rice fields, the damage can be direct or indirect: destruction of plants cut and torn up in whole clumps, digging of burrows in dikes, bunds and canals, etc. irrigation, causing gullies and water diversions requiring urgent and costly repairs (Heinrichs and Barrion, 2004). Apart from insects, birds and rodents, nematodes attack the roots to a much greater extent, affecting the circulation of sap in the plant. The result is a reduction in the root system, vigour, tillering and stem height.

12.3. Rice diseases

Rice is affected by a number of viral, fungal and bacterial diseases.

12.3.1. Fungal diseases

Rice blast is the main fungal disease of rice, also known as leaf blast, caused by Magnaporte grisea. This disease has been reported in almost all rice-producing countries around the world and can infest rice plants at all stages of development, but especially in the nursery and at flowering. Its importance depends on the presence of inoculum (which varies from region to region). The spread of this disease is favoured by drought, high humidity and high doses of nitrogen (Agarwal et al., 1994). Symptoms are observed on the leaves (initially appearing as whitish or greyish spots, which rapidly enlarge in the presence of favourable conditions (water deficit, dew, high nitrogen content in the soil (Agarwal et al., 1994)); the nodes of the stems, the various parts of the panicle, as well as the grains.

The use of fungicides and resistant varieties are the main strategies used to combat this disease. Other fungal diseases (helminthosporiosis, rhyncosporiosis, cercosporiosis, sheath wilt caused by Rhizoctonia solani) rarely cause significant damage.

12.3.2. Bacterial diseases

The main one is bacterial wilt (Xanthomonas campestrispv. Oryzae), which occurs mainly in Asia. This is followed by streak rot (Xanthomonas translucens) and brown sheath rot (Pseudomonas ovaginae); the latter has recently been identified in Asia. There is also helminthosporiosis (Helminthosporium oryzae or Cochliobolus miyabeanus) and rice pellicular blight (Rhizoctonia solani) (WARDA, 2000). Control of these diseases requires the use of resistant varieties and bactericides.

12.3.3. Viral diseases

According to the West Africa Rice Development Association (WARDA) in 2000, rice yellow mottle virus (RYMV) is a viral plant disease. According to Banwo et al (2001), the abstract rice yellow mottle virus (RYMV) belongs to the Sobemovirus group and causes the only known disease of rice (Oryza sativa) virus specific to the African continent so far. According to Sadou et al. 2008, Yellow Mottle Virus (RYMV) is the main viral disease affecting irrigated rice cultivation in Africa. It is transmitted by insect vectors through lesions or mechanically during crop maintenance, for example by hoeing during weeding. Also of note is the tungro virus, which limits rice production in Bangladesh, India (1967), Indonesia, Malaysia, the Philippines and Thailand, with yield losses of up to 100% if no control measures are taken.

13. Disease and insect control methods

Several methods can be used to combat pests and diseases, including: cultural control, biological control, chemical control, genetic control, physical control and integrated control.

13.1. Cultivation practices

These are cultivation practices that prevent the multiplication of harmful insects and at the same time favour the growth and development of rice. One example is weeding, which destroys certain weeds that serve as a refuge or biotope for insect pests (Brenière, 1969).

13.2. Biological control

The aim is to limit the density of pests through action of natural enemies such as predators, parasites or pathogens, as well as autocides (using sterile males) and confusion control (chemical communication molecules) (Siaussat, 2005). Two strategies can be distinguished: conservation and increasing natural enemies. Conservation simply means not destroying natural enemies through cultivation practices and other interventions such as chemical control. Augmentation, on the other hand, involves artificially increasing the number of natural enemies in the environment.

13.3. Genetic control

Genetic control consists of using tolerant varieties; it is generally inexpensive for farmers and not very restrictive for the environment. Varieties can be distinguished according to their resistance and classified as low, medium and high resistance (Brenière, 1969).

13.4. Chemical control

Chemical control using insecticides has the advantage of generally being fairly effective, killing the insect in a short space of time. In addition, insecticides are often widely available in shops. However, the use of insecticides is fairly expensive and poses technical constraints. According to Brenière (1962), chemical control is in fact the only fairly rapid and effective method available to us.

13.5. Physical combat

Physical pest control is a method of pest control that involves the direct elimination or hunting of pests on crops. It uses scarecrows, nets, abrasive particles, temperatures (heat or cold), ionising radiation, gases, statutes and sounds to frighten or chase pests away from crops.

13.6. Integrated pest management

Integrated pest management involves keeping insect pests at a sufficiently low level without disrupting the stability of the ecosystem, while ensuring the economic efficiency of the rice crop. Integrated pest management involves reasoning and combining the different control techniques. According to the Organisation Internationale de Lutte Biologique (O.I.L.B, 1973 in Ferron, 1999) quoted by Johanna Villenave in 2006, integrated protection is a concept of plant protection based on knowledge of the environment and the dynamics of the populations present in agroecosystems.

CONCLUSION

Rice (Oryza sativa L.) is the staple food of more than half the world's population (Guigaz, 2002). However, this crop faces a number of obstacles to production, including attacks by insect pests. The use of synthetic insecticides to combat these pests has a negative impact on the biodiversity of ecosystems and on human health, yet nature provides us with biological means of combating these insects by means of crop phytoprotection, such as aqueous neem seed extract.

BIBLIOGRAPHICAL REFERENCES

Abdoul A., Hughes J. & Diallo A. (2001). Yellow mottle in rice. Economic importance, diagnosis and management strategies. WARDA Proceedings 1: 250 - 251.

Abdoulaye s, Abou, Togola, & Ali, Toure. 2013. Proposition pour une optimisation des performances a la riziculture en Afrique de l'ouest Senegal: 34pp.

Abo M. E., Alegbejo M. D., Sy A. A. & Sere Y. (2004). Retention and transmission of Rice yellow mottle virus (RYMV) by beetle vectors in Cote d'Ivoire. Revue de l'Association ivoirienne des sciences agronomiques 16 : 71 - 76.

Adégbidi, A. (1994). Cours de gestion des exploitations agricoles. FSA/UNB; p81.

Adégbidi, A. (2018). Socio-economic determinants of the adoption of new technologies: The case of maize in the department of Atacora in the Republic of Benin. Postgraduate doctoral thesis in economics (rural economics), CIRES, University Côte d'Ivoire, Abidjan, 169p.

Adégbola, P. and Midingoyi, S. (2004). Efficacité technico-économique des agriculteurs et rizicultrices du Centre-Benin. Paper presented at the NATIONAL 2004 scientific workshop.

Adégbola, P. & Sodjinou, E. (2017). Study of the rice sector in Benin. Final report; p 231. ADRAO. (1990). WARDA Annual Report, 1990. 1: 2-36. WARDA. (2000). Highlights of Activities. WARDA Annual Report, 2000. 1 : 1-12. WARDA. (2010). Insect pests ofrice in west Africa WARDA, Bouaké (Côte - d'Ivoire). 73 p

WARDA. (2014). NERICA rice variety as a source of life in West Africa. Bouaké, Ivory Coast: 64pp.

WARDA. (2016). Annual Report. WARDA, Bouaké (Côte - d'Ivoire). 59 p.

WARDA. (2017). New Rice for Africa. www.warda.org.

Agarwal P. C., Mortensen N. C. & Mathur B. S. (1994). Seed-borne diseases of rice and phytosanitary tests. WARDA. 80 p.

Ahoyo, R.N.A. (2018). Economics of production systems integrating rice cultivation in South Benin: potentialities, constraints and perspectives. PhD thesis, Peter Lang, 270 pages.

Akinsola E. & Agyen - Sampong M. (2014). The ecology, bionomics and control of rice stem- borer in West Africa. Insect Sei Applic. 5 (2): 69 - 77.

Aminatou B. (2009). Inventaire des phytoinsecticides pour la protection des grains au cours stockage contre les ravageurs dans la zone sahélienne (cas de l'Extreme Nord du Cameroun). Final report, Réseau Anafe/Raft-AHT. 32p.

Amuli (2015). Comparison of two intermittent water management techniques on the yield of five irrigated rice varieties at Kiringye in the Ruzizi plain, unpublished dissertation, UCB.

Anonymous (2010). Mémento de l'agronome, P.691 - 1254 éd. Quae, Montpellier - France.

Arhent C., Cramer H. H., Mogk M. & Peshel H. (1983). Economic impact of crop losses. 10th International Congress of Plant Protection. November 10- 25, 1983. 1: 65 - 73.

Arraudeau M. (1998). Irrigated rice. Maisonneuve & Larousse, Paris. 656 p.

Arraudeau, M.A. and Vergara, B.S. (2018). A farmer's primer on growing upland rice. Los Baños, Philippines, IRRI. 95 pages.

Attikou, A. (2014). Suivi de la fertilité des sols des aménagements hydro-agricoles : Cas des périmètres rizicoles de Bonféba et Lata. Niamey, Office National des Aménagements Hydro- agricoles. 39pp. Akintayo I., Cissé B., and L. D. Zadji. 2008. Guide pratique de la culture des NERICA de plateau. Cotonou, Benin: Africa Rice Center (WARDA), 36pp.

Awinakai J. (1998). Rapport de stage de découverte d'un milieu humain. University of Dschang. 42 p.

Awoderu V. A. (1991). The Rice Yellow Mottle virus situation in west Africa. Tropical pest management 37: 356-362.

Bamba, B. (2015). Contribution to the evaluation of losses due to rice blast. Dissertation, University of Ouagadougou. 51 p.

Barker, R. & Herdt, R.W. (2016). Rainfed lowland rice as a research priority - an economist's view. In Rainfed lowland rice, pp. 3-50. Los Baños, Philippines, IRRI.

Beauvilain A. (2011). Notes on the cities of the northern province. Revue de Géographie du Cameroun, 2 (1): 25-32.

Beckage N.E. (1985). Endocrine interactions between endoparasite insects and their hosts. Annual Revue of Entomology 30: 371-413.

Bénech V., Quensière J., Vidy G., (2012). Hydrology and physicochemistry of North Cameroon floodplain waters. Cah. Orstom, ser. Hydrol, 19 (1): 15-36.

Bessong à Beyeck, L. (2006). Potential trade between Cameroon and its border partners. Mémoire d'Ingénieur d'Application de la Statistique, option gestion. Institut Sous Régional de Statistique et d'Économie Appliquée (I.S.S.E.A).

Bhan, V.M. (2016). Relation between growth and yield characters in upland paddy. Riso, 15: 229-232.

Bhan, V.M. (2018). Influence of row spacing on upland paddy. Riso, 17(2): 149-154.

Blache J., Miton F. (2011). Première contribution à la connaissance de la pêche dans le bassin hydrographique Logone-Chari-lac Tchad. Paris, ORSTOM, Mémoire n° 4, 143 p.

Billon B., Bouchardeau A., Roche M., Rodier J. (2015). Monographie hydrologique du Logone. Paris, Orstom, 8 vols, 770 p.

Biscaldi R. (2014). Hydrology of the Logone-Chari-Chad water table. BRGM, Report 70-Yao-003.

Biswas, A. K., Islam, A., Nayek, B. & Choudhuri, M.A. (2017). Water - stress induced susceptibility to pests in rice plants (Oryza sativa). Indian Biologist 14 (1): 13 - 20.

Bouafia A, 2002, La rousse agricole: 755 pp.

Bouchardeau A. (2015). Monographie hydrologique du Logone. Paris, Orstom, 8 vols, 770 p.

Bouman, B. A. M. & Tuong P. (2011). Field water management to save water and increase it productivity in irrigated rice agriculture water management 49 (1), 11-30.

CBLT. (2016). Studies of water resources in the Lake Chad Basin, with a view to a development programme. In: Groundwater resources in the Lake Chad Basin, LCBC, Cameroon, Niger, Nigeria, Chad, UNDP, Rome FAO.

Cessouma A. (2008). L'agriculture irriguée et perspective et l'économie d'eau au Mali 67pp.

CIRAD-GRET, (2009). MEMENTO de l'agronome, New edition, Ministère des affaires étrangères (France) Pp799-811.

Christian Seignobos & Olivier Iyébi-Mandjek, (2005). Atlas de la province Extrême-Nord Cameroun, Paris, IRD Éditions, 171 p

CTA. (1995). Integrating biological control and host plant resistance. Proceedings of a CTA/IAR/IIBC seminar. Addis Ababa, Ethiopia, 9-14 October 1995. 383 p.

Dakouo, D., Nacro, S. (2018). Development of a rational insect pest management system on SEMRY rice fields. Insect. Sc. Appl. 15 (5/6): 565- 570.

De Datta, S.K. & Feur, R. (2017). Soils on which upland rice is grown. In Major research in upland rice, pp. 27-39. Los Baños, Philippines, IRRI.

De Datta, S.K. & Vergara, B.S. (2017). Climate of upland rice regions. In Major research in upland rice, pp. 14-26. Los Baños, Philippines, IRRI.

Decraene P. (2017). Lettre de Maroua: l'ombre portée du Nigeria. Le Monde, 29 to 29 July 2014: 8.

De Groot, I. (1996). Protection of stored leguminous cereals. CTA. Agrodoc18: 3-45. Delseny M. (2016). Rice, a model in genomics. Biofutur 203, 30-35.

Delvare G. & Aberlenc P. 1989. Insects of Africa and Tropical America. Clé pour la reconnaissance des familles. 297 p.

Denham M. (2014). Voyages et découvertes dans le nord et les parties centrales de l'Afrique, exécutés pendant les années 1822, 1823 et 1824. Trad. Eyries et de la Renaudiere, Paris, A. Bertrand, 3 vols. 367 p., 379 p., 428 p.

Diaz, OL (2009). Climate change, poverty and transport in cities south of the Sahara. Université Cheick Anta Diop, 310p.

Dobelmainn J.P. (2010). Riziculture pratique. Riz pluvial. Techniques vivante. Agency for Cultural and Technical Cooperation. Presses universitaires de France, Paris; France.

Dubreuil P. (2015). Étude des inondes sur un petit bassin de la région de Maroua, Nord- Cameroun. Paris, ORSTOM, Annuaire hydrologique de la France d'Outre-Mer: 15-27.

Durand J. R. & Lévêque C. (2015). Aquatic flora and fauna Sahelo-Sudanian Africa. Paris, Orstom, IDT 45, 484 p.

Engola-Oyep J. (1991). Assurer la survie des périmètres hydro rizicoles à l'heure l'ajustement structurel. Cahier des Sciences Humaines 27: 53-63

WARDA Spirit. (2004). The Africa Rice Center. WARDA Journal 5: 2 - 4.

FAO. (2000). Rice. FAO. Statistical Bulletin 1: 20-73.

FAO. (2001). Statistical Yearbook; www.fao.org. 31/03/2017. FAO. (2002). Statistical yearbook of production 56 :175-176.

FAO. (2004). Extraordinary Summit of Heads of State and Government of the African Union on Employment and Poverty Reduction in Africa. Situation of agriculture and rural populations in Africa.

FAO. (2011). The state of food insecurity in the world. How international price volatility affects the economy and food security of countries. Rome 62 p.

FAO. (2013). Rice. FAO. Statistical Bulletin 2: 2-10.

FAO. (2014). FAO. Statistical Bulletin 1: 24-54.

FAO. (2017). Improved rainfed rice farming systems, Rome.

FAO. (2018). Statistical Yearbook; www.fao.org.

Fritsch P., (2013). Geological aspects of the floodplains of North Cameroon.

Guigaz M. (2000). Mémento de l'agronome. CIRAD GRET, Paris. 1690 p.

Guillard J. (2015). Studies and work on fishing in the Cameroonian Logone Basin. Garoua, Archiv. de l'Inspection forestière du Nord.

Guiscafre J. (2016). Influence des aménagements anti-érosifs sur l'écoulement des mayos Kapsiki - Bassins versants de Mogodé. Yaounde, Orstom-Ircam, 40 p., graphs.

Gupta, P.C. & O'Toole, J.C. (2018). Upland rice - global perspective. Los Baños, Philippines, IRRI. p. 1; 63-68; 306-310.

Hamadoun A. & Traoré (1997). Status of rice mottle in Mali: the case of the Office du Niger. Pp 20-22 In: International Symposium Rice Yellow Mottle Virus (18-22 September 1997). M'bé/ Bouaké, Ivory Coast.

Hanks, J. (2006). Mitigation of human-elephant conflict in the Kavango-Zambezi Transfrontier Conservation Area, through Community Based Problem

Animal Control, with particular refrence to the use of chilli peppers. Report prepared by Conservation International.

Hirsch, R. (2018). African rice farming: importance and issues (rice and rice policies in West Africa and the PSI/CORAF zone). In Pour un développement durable de l'agriculture irriguée dans la zone Soudano-Sahélienne : Synthèse des résultats du Pôle régional de Recherche sur les systèmes irrigués (PSI/CORAF), Proceedings of the Seminar Dakar (Senegal) from 30 November to 3 December 1999, CIRAD/CF/CTA.

Hugon Ph. (2014). Agricultural sectors and macroeconomic policies. In Benoit-Cattin M, Griffon M, Guillaumont P, éditions économie des politiques agricoles dans les pays en développement, les aspects macroéconomiques, Paris, revue française d'économie.

Huke R.E. & Huke E.H. (1990). Rice. Then and now. Manila, International Rice Research Institute 1: 1- 44.

IITA (International Institute of Tropical Agriculture). (1987). Annual report for 1987. IITA, Ibadan. 160 p.

IITA (International Institute of Tropical Agriculture). (2000). Annual Report and Research Update 2000. IITA, Ibadan. 145 p.

IITA (International Institute of Tropical Agriculture). (2013). Annual report and research update 2013. IITA, Ibadan. 192 p.

Ingram, W. R. (2016). Biological control of graminaceous stem-borers and legume podborers. Insect. Sei. Appl, 4 (1/2). Pp 205-209.

IRRI (International Rice Research Institute). (1997). Annual report. IRRI, Manila. 90 p.

IRRI (International Rice Research Institute). (2015). Problems in rice cultivation. Identification guide second ed. Manila Philippines. Pp :71-79.

IRRI. (2017). Rice facts. Los Baños, Philippines. Manila. 87 p.

IRRI. (2018). International upland rice genetic survey. Los Baños, Philippines. Manila. 45 p.

Issoufou, A. & Issa, G. (2006). Détermination des surfaces et estimation de la production de riz hors aménagements hydro-agricole. Study report. Niamey, Programme d'Appui à la Filière Riz. 43 pp.

Johnson, D.E. (2017). Les adventices en riziculture en Afrique de J'Ouest, WARDA, Bouaké (Cote -d'Ivoire), PP: 1-2.

Kaung V. T., John & Alam. (1985). Rice production in Africa: an over view, in Rice Improvement Eastern, central and southern Africa. Pp 7-27 In: Proceeding of the international Rice workshop. International Rice Research Institute, Philippines.

Matsushima S. (2016). Crop Science in rice, theory of yield determination and its application. Fuji Publishing Co, Ltd, edit.3.

Migabo, K. (2009). Essai d'introduction de six variétés de riz irrigué dans les conditions edapho climatiques de la plaine de la Ruzizi. Unpublished UCB dissertation: 51 pp.

Misra, R.K. Mathur, S.C. & Chaudhary, R.C. (2016). Breeding rice varieties resistant to diseases. In National Symposium on Increasing Rice Yield in Kharif, pp. 95-114. Cuttack, India, Central Rice Research Institute.

Monty P. Jones. (2016). The rice plant and its environment. ADRA0 training guide, Bouaké (Cote d'Ivoire), n034: 28 p.

Moormann, F.R. (2016). General assessment of land on which rice is grown in West Africa. Paper presented at the Seminar on Rice Soil Fertility and Fertilizer Use, 22-27 January 2016. Monrovia, Liberia, West Africa Rice Development Association.

Mukhopadhya, S.K., Ghosh B.S. & Maity, H. (2014). Weed problem in upland rice and approaches to solve the problem by use of new herbicides. Oryza, 8: 269-274.

Niyondagara, T. (1984). Contribution à la sélection du riz pour un environnement pluvial. Dissertation, FACAGRO, U.B, Bujumbura: 112pp.

Noëlle Terpend (2017). A practical guide to the value chain approach. The case food supply and distribution in cities. Collection "aliments dans les villes", 39 pages.

Nwilene F. E. (2000). Highlights of WARDA Activities. WARDA Annual Report 2000, WARDA, Bouaké. 30 p.

Paul Ondo Ovono, Thaddée Gatarasi, Daniel Obame Minko, Drice Miyoumbi Koumagoye & Claire Kevers (2014). Study of insect population dynamics on NERICA rice crop under Masuku conditions, South-East Gabon (Franceville). Int. J. Biol. Chem. Sci. 8 (1): 218-236, February 2014. ISSN 1997-342X (Online), ISSN 1991-8631 (Print).

Rao, M.V. (2016). Pushing up rice yields in uplands. Intensive Agric, 16: 4-7.

Rao, C.N. & Murty, K.S. (2015). Effect of split application of nitrogen on tiller survival and yield in direct seeded upland rice. Indian J. Agric. Sci. 45: 183-185.

Rosenfeld M.F. (2017). Estimating crop yields using the sample cut method. In : Économie rurale. N°18. pp. 25-28.

Roupsard M. (2015). North Cameroon, openness and development. Coutances, Impr. C. Bellee, 516 p.

Roy, J.K., Israel, P. & Panwar, M.S. (2016). Breeding for insect resistance for rice. Oryza, 6: 38-44.

Sadou I., Woin N., Ghogomu T.R. and Hamasselbe A. (2013). Inventory of predators and parasitoids of insect pests of rice in the far north region of Cameroon. GBS Publishers & Distributors (India), Indian Journal of Applied

Agriculture Research. http: //www. gbspublisher. com/Volume/ijaarv1n1.htm.

Sadou I., Woin N., Ghogomu T.R. and Hamasselbe A. (2013). The inventory of insect pests and vectors of rice yellow mottle virus in rice ecosystems in the region of the Far North of Cameroon. GBS Publishers & Distributors (India), Indian Journal of Applied Agriculture Research. http://www.gbspublisher.com/Volume/ijaarv1n1.htm.

Sadou I. (2011). Inventory of insect vectors of rice yellow mottle virus rice ecosystems in the Far North of Cameroon. Book. Editions universitaires européennes, 2011, 84 p. ISBN: 978-613-1-58823-5. Schaltungsdienst Lange, Berlin, Germany. www.thebookedition.com /inventaire - des - insectes - vecteurs - sadou- ismael - p-66593. html or www.ecochimp.com/ inventaire-des-insectes-vecteurs-du-virus-de-la- panachure-jaune-du-riz-sadou-ismael--p-36750.html.

Sadou I., Woin N., Ghogomu T.R. & Djonmaila (2008). Inventory of insect pests and vectors of rice yellow mottle in the irrigated areas of Maga (Far North). Tropicultura, Vol. 26, No. 2, 2008, pp. 84-88.

Schalbroeck, J.J. (2001). Rice: 85-105 In: Raemakers, R.JH. (Ed): Agriculture en Afrique Tropical. D.G.C.I., Rue de petits carmes, Brussels, Belgium :1634PP.

Schalbroeck J.J. (2001). Cereal plants. Agriculture in Tropical Africa 1: 10-169.

Schalbroeck J. J. (2016). Rice. In: Agriculture in Tropical Africa. Ministry of Foreign Affairs of Foreign Trade and International Cooperation, Brussels, Belgium, p. 99 - 105.

Sebillotte M. (2016). Agronomy and agriculture. Essai d'analyse des taches de l'agronome, Cah. ORSTOM, ser. Biol, (24).

Seignobos C. (2015). L'arbre et la cite dans la zone soudano-sahélienne (les exemples du Tchad et du Cameroun septentrional). Yaounde, Rev. de Géogr. du

Cameroun, 2 (1): 49-52.

Sere, Y. & Eble S. (2017). Rice production and rice research in Burkina Faso. Report presented at the session of the international rice commission held in GOIANA, Brazil, 4-9 February 1990. 25 p.

Sie M., Kabore K. B., Dakouo D., Dembele Y., Sedga Z., Bado B. V., Ouedraogo M., Moukoumbi Y. D., Ban. M. & A. Traore. (2006). Fiches techniques des quatre nouvelles variétés de riz de type NERICA pour la riziculture de bas-fond / irriguée au Burkina Faso : FKR 56 N, FKR 58 N, FKR 60 N, FKR 62 N. INERA, Bobo Dioulasso. Burkina Faso, 5 p.

Singh, R.K. & Singh, C.V. (2016). Central Rainfed Upland Rice Research Station annual report 2016. Hazaribagh, India, ICAR.

Singh, G.R, Parihar S.S. & Chaure (2017). Response of organic manures in a rice (Oryza sativa) - Chickpea (Cicer arietinum) crop sequence. Taxonomy of the genus Oryza (Poaceae): International Rice Research Newsletter 24 (3): 24 - 25.

Sukmana, S. & Suwardjo, H. (2017). Prospects of vegetative soil conservation measures for sustainable upland agriculture. IARD J., 13 (1-2): 1-7.

Tran, D.V. (2016). Overview of upland rice in the world. In Progress in Upland Rice Research, pp. 51-66. Los Baños, Philippines, IRRI.

United States Department of Agriculture (USDA). (2014). National agricultural statistics service agricultural statistics 2014, united states government printing office washington: 2014

Vaughan, D. A., Morishima, H. & Kadowaki, K. (2003). Diversity in the Oryza genus. Current Opinion in Plant Biology, 6: 139-146.

Vermura, Y. & Mayasaka, A. (2018). Causes of rice yield decrease resulting from continual direct sowing culture and its relation to phosphorus nutrition.

Proc. Crop Sci. Soc. Japan, 42 : 116-122.

WARDA (2006). Progress Report 2003-2005 joint interspecific hybridization between African and Asian Rice species (Oryza glaberrima and O. sativa), P. 151. Cotonou, Benin.

Zadia, H, C. (2016). Analyse de la compétitive dans la gestion de la diversité variétale de riz dans la plaine de la Ruzizi (cas de lubarika, luberizi, et kiringye), unpublished dissertation UCB: 40pp.

Zhang, Y. (2004). Factors affecting chines farms decision to adapt a water saving technology. Can.j. agric. econ. 56, 61pp.

Printed by Books on Demand GmbH, Norderstedt / Germany